ISBN 978-0-364-77857-9
PIBN 11271129

This book is a reproduction of an important historical work. Forgotten Books uses
state-of-the-art technology to digitally reconstruct the work, preserving the original format
whilst repairing imperfections present in the aged copy. In rare cases, an imperfection in
the original, such as a blemish or missing page, may be replicated in our edition. We do,
however, repair the vast majority of imperfections successfully; any imperfections that
remain are intentionally left to preserve the state of such historical works.

For support please visit www.forgottenbooks.com

Historic, Archive Document

Do not assume content reflects current
scientific knowledge, policies, or practices.

Seasonal Trends in the Nutritive Content of Important Range Forage Species Near Silver Lake, Oregon

O. Eugene Hickman

ABSTRACT

Moisture, crude protein, ash, calcium, phosphorus, calcium:phosphorus ratios, crude fiber, crude fat, and apparent digestibility were determined for 15 species in three vegetal classes on seven vegetation types. These measurements were contrasted by classes, species, and type, and ungulate management inferences were drawn.

KEYWORDS: Nutrient regime (animal), range production, forage plants.

CONTENTS

ACKNOWLEDGMENT

This study was conducted while the author was a graduate student at Oregon State University. Financial assistance was provided by the U.S. Forest Service, Pacific Northwest Forest and Range Experiment Station, Portland, Oregon.

The author is indebted to his former major professor, Dr. D. W. Hedrick, for much needed advice and guidance in planning and implementing this project. Sincere thanks are extended to J. Edward Dealy of the Experiment Station, Dr. J. E. Oldfield of the Oregon State University Animal Science Department, and Allen Bond, Oregon State University graduate student, for their part in assisting and guiding the author in field and laboratory work.

INTRODUCTION

Livestock and game managers are increasingly aware of the nutritional aspects of range management. An adequate diet for range animals is essential for the efficient production of animal products.

Deficiencies in protein, phosphorus, and carotene (vitamin A) are common on western ranges (Cook and Harris 1968) and have led to studies on nutritional values of forage plants (Cook and Harris 1950a, Dietz et al. 1962, Beath and Hamilton 1952, Skovlin 1967, Vallentine and Young 1959). Such information can be used to estimate the seasonal adequacy of animal diets and is needed in order to adopt sound management practices.

A limited number of such studies have been undertaken in central Oregon by scientists of the Pacific Northwest Forest and Range Experiment Station (Dealy 1966), at the Squaw Butte Experiment Station near Burns, Oregon (Raleigh and Wallace 1965, Wallace et al. 1961), and at Oregon State University (Smith 1963, Urness 1966).

The Silver Lake mule deer winter range, 80 miles south of Bend, is important from the standpoint of big game use and livestock production. Extensive research in livestock and big game management has been conducted there but included little nutritional evaluation of native forage. Studies have included effects of big sagebrush (*Artemisia tridentata*) in deer diets (Smith 1963) and determination of crude protein levels in native grass and shrub species during winter and early spring (Urness 1966). This study was developed to:
1. Define seasonal trends for the crude protein, fiber, and fat (ether extract), moisture, ash (total mineral matter), calcium, phosphorus, and apparent digestibility in selected forage species; and
2. Determine if these trends vary between ecologically different vegetation types.

THE STUDY AREA

The study area is 6 miles southwest of the town of Silver Lake, Oregon, in Lake County, elevation approximately 4,700 feet. It is in the forest fringe adjacent to the High Desert and typical of the Silver Lake deer winter range. Here, the Pacific Northwest Forest and Range Experiment Station and the Oregon Wildlife Commission established three 170-acre enclosures that were sampled in this study.

Climate

The High Desert is dry, with hot summers and cold winters. Most precipitation occurs as snow, although significant rain may occur in May and June (Urness 1966). The growing season is about 79 days, with frost a possibility during any month.[1]

Climatological data were obtained from the weather station at Silver Lake and are compared to 26 years of climatic records (fig. 1).

The average annual precipitation at Silver Lake is 9.9 inches (Johnsgard 1963, p. 80), which is probably lower than that at the study plots. During December 1964, midway in the study, precipitation totaled 7.68 inches, far above the average.

Soils

Soils developed from volcanic material, including tuff and tuff breccia, hard basalt, and pumice, have been mapped and described by Herman (see footnote 1).

---

[1] L. D. Herman. Soil survey report for Silver Lake Experimental Range study area, Lake County, Oregon. Unpublished manuscript on file at Pacific Northwest Forest and Range Experiment Station, La Grande, Oreg., 18 p., 1965.

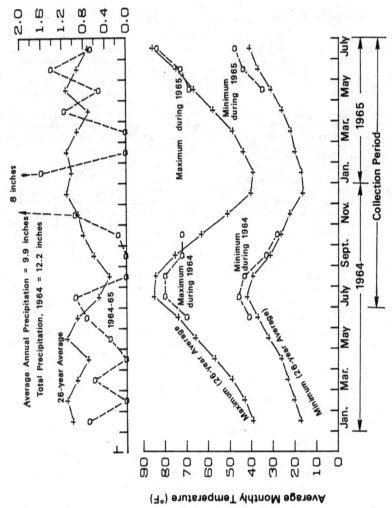

Figure 1.--Temperature and precipitation data for Silver Lake weather station.

Vegetation Types

Vegetation types within the enclosures were delineated and described by Driscoll and Dealy[2] and may be descriptively titled as:

1. *Pinus ponderosa/Purshia tridentata/Festuca idahoensis*
2. *Pinus ponderosa/Cercocarpus ledifolius/Festuca idahoensis*
3. *Purshia tridentata-Artemisia arbuscula/Poa secunda*
4. *Purshia tridentata/Festuca idahoensis*
5. *Artemisia arbuscula/Agropyron spicatum*
6. *Artemisia arbuscula/Poa secunda.*

An area studied just outside the enclosures was dominated by western juniper (*Juniperus occidentalis*), with an understory primarily of Idaho fescue (*Festuca idahoensis*), and was named:

7. *Juniperus occidentalis/ Festuca idahoensis.*

Livestock and Game Use

The mule deer herd moves to this winter range in mid-December and leaves in mid-April. A small herd is resident year-round.

Livestock are regulated by cow-calf permits issued by the U.S. Forest Service. The grazing season was from late May until the beginning of July.

---

[2] R. S. Driscoll and J. E. Dealy. A study to investigate the foraging habits of mule deer and cattle and factors affecting them on the Silver Lake herd range. Unpublished manuscript on file at Pacific Northwest Forest and Range Experiment Station, La Grande, Oreg., 27 p., 1964.

METHODS AND PROCEDURES
Species Selection

The species studied are shown in table 1. Grasses and browse species were selected based on their abundance and importance as forage for deer and/or livestock.

Forbs were less plentiful, so emphasis was on abundance. Mountain lily (*Leucocrinum montanum*) had considerable use, probably by deer. The importance of the other forbs as forage is uncertain. They are representative of forbs available to deer and/or cattle during the spring and summer.

Sampling Dates and Techniques

The study period, July 1964 through July 1965, included 15 sampling dates and was planned to include one complete annual cycle for the species selected. Consequently, sampling was scheduled by growth stages of important species.

All species were collected by vegetation types (table 1). Several species were important in more than one type: bitter brush (*Purshia tridentata*), squirreltail (*Sitanion hystrix*), and Idaho fescue. Ecotones were avoided in the collections.

Only one species, Idaho fescue, was collected from the *Juniperus occidentalis/Festuca idahoensis* type. Idaho fescue grows vigorously around the base of juniper trees. Although accessible to livestock, it is sometimes lightly used or untouched. Idaho fescue was collected from these micro-sites and compared with collections of the same species from other types.

Clipping was done to approximate grazing and included only current annual growth except for mountain mahogany (*Cercocarpus ledifolius*), which included previous years' leaves.

Table 1.--*Species selected for sampling (*) and other plants referred to in this report (scientific nomenclature follows Hitchcock (1950) for grasses and Peck (1961) for other species)*

Scientific name	Common name	Vegetation type where sampled
Grasses and Grasslike		
Agropyron spicatum (Pursh) Scribn. and Smith	Bluebunch wheatgrass*	*Artemisia arbuscula/Agropyron spicatum*
Carex Rossii Boott	Ross' sedge*	*Pinus ponderosa/Purshia tridentata/Festuca idahoensis*
Festuca idahoensis Elm.	Idaho fescue*	*Pinus ponderosa/Purshia tridentata/Festuca idahoensis*
Koeleria cristata (L.) Pers.	Junegrass*	*Artemisia arbuscula/Agropyron spicatum*
Poa secunda Presl.	Sandberg bluegrass*	*Artemisia arbuscula/Agropyron spicatum*
Sitanion hystrix J. G. Smith	Squirreltail*	*Purshia tridentata-Artemisia arbuscula/Poa secunda*
Stipa thurberiana Piper	Thurber needlegrass*	*Purshia tridentata-Artemisia arbuscula/Poa secunda*
Forbs		
Delphinium Nuttallianum Pritz.	Upland larkspur*	*Purshia tridentata/Festuca idahoensis*
Leucocrinum montanum Nutt.	Mountain lily*	*Artemisia arbuscula/Poa secunda*
Senecio integerrimus Nutt. var. *exaltatus* (Nutt.) Cron.	Tall western senecio*	*Pinus ponderosa/Purshia tridentata/Festuca idahoensis*
Sidalcea oregana (Nutt.) Gray	Oregon sidalcea*	*Purshia tridentata/Festuca idahoensis*
Shrubs		
Artemisia arbuscula Nutt.	Scabland sagebrush*	*Artemisia arbuscula/Poa secunda*
Artemisia tridentata Nutt.	Big sagebrush*	*Purshia tridentata/Festuca idahoensis*
Cercocarpus ledifolius Hook.	Mountain mahogany*	*Pinus ponderosa/Cercocarpus ledifolius/Festuca idahoensis*
Juniperus occidentalis Hook.	Western juniper	--
Pinus ponderosa Dougl.	Ponderosa pine	--
Purshia tridentata (Pursh) DC.	Bitter brush*	*Pinus ponderosa/Purshia tridentata/Festuca idahoensis*

Composite samples (per observation) were made by collecting small amounts of material from plants (20-40) that were well distributed through each vegetation type. Subsequent samples taken at later sampling dates were from the same general area in each type but not necessarily from the same individual plants. Average growth stage was estimated for each composite sample.

All material was placed in airtight plastic bags after collection and taken to the laboratory (1-2 days later) for weighing and subsequent chemical analyses.

Methods of Analysis

Samples were analyzed for crude protein, fiber, and fat (ether extract) and for ash (total mineral matter) by methods approved by the Association of Official Agricultural Chemists (1965).

Samples analyzed for calcium and phosphorus were dry ashed.[3/] Calcium was determined by atomic absorption spectroscopy (Willis 1960) and phosphorus by the Vanadomolybdophosphoric yellow color method in a sulfuric acid system (Jackson 1958).

Apparent digestibility (dry matter disappearance) was determined by the artificial rumen technique of Smith (1963) and Bedell (1966). Rumen juice was collected at midday from a fistulated steer having free access to grass hay and water and then used in a 24-hour digestion period. One 48-hour period was utilized for a comparison of these two treatments.

[3/] Personal communication with Dr. O. C. Compton, Oregon State University, Corvallis.

RESULTS AND DISCUSSION

The data have been summarized graphically for several species to give a picture of seasonal trends and relationships among constituents.

Nutrient content is only one of several criteria that should be used to judge the value of a forage plant. Data presented here will be most useful when related to other important factors such as species abundance, season of use, and palatability.

Nutrient Relationships by Vegetal Classes

GRASSES AND SEDGES

Calcium, phosphorus, crude protein, apparent digestibility, and moisture are associated in the grass and sedge species. Trends were similar throughout the season (figs. 2 and 3). An early spring increase in moisture was accompanied by similar increases in other constituents. Maximum moisture content occurred before the boot stage. This was followed by a downward trend with nutrients leveling off later in the season. Results are similar to those of Skovlin (1967) and Whitman et al. (1951).

Gordon and Sampson (1939) found a continuous and orderly decline in crude protein, calcium, and phosphorus from the earliest appearance of leaf blades to maturity in grasses. Wallace and Raleigh (1962) found that both apparent digestibility and crude protein of crested wheatgrass (*Agropyron aristatum*) decreased significantly as the forage matured. The trend in apparent digestibility reported here for Idaho fescue is similar to that of Pearson (1964) for Arizona fescue (*Festuca arizonica*), although absolute values are greater.

Moisture content remained high from early leaf stage until anthesis (flowering) in the grasses and from early leaf stage until seed formation in the sedge. After these stages were reached, moisture decreased rapidly. The rate of loss of other constituents were inverse to this. In general, calcium, phosphorus, crude protein, and apparent digestibility decreased at a rapid rate when moisture loss was minimal. Gordon and Sampson (1939) reported that the most rapid change in nutrient content took place between the periods of leaf development and full bloom. When moisture began to decrease at a faster rate, the other constituents tended to level off or slow their rate of decrease, Sandberg bluegrass (*Poa secunda*) excepted. For this species, apparent digestibility decreased as rapidly as moisture.

Although calcium and phosphorus decreased with plant maturity, the ratio of calcium to phosphorus widened as calcium decreased at a slower rate than phosphorus. Compared with moisture content, the calcium:phosphorus ratio was narrow when the moisture content was high and widest when moisture content was low.

Seasonal trends of crude fiber and ash were opposite to those for calcium, phosphorus, crude protein, apparent digestibility, and moisture. Crude fiber and ash increased during the season at roughly parallel rates (except for ash content decline in some species during early summer). Gordon and Sampson (1939) found that crude fiber was lowest in grasses during early plant growth and highest at maturity. Vallentine and Young (1959) showed crude fiber negatively related to crude protein content in grasses.

Crude fat content of grasses showed no relationship with other nutrients except for a slight but temporary drop in both crude fat and ash in some species at anthesis. This is consistent with studies of Kentucky bluegrass (*Poa pratensis*) where plant development showed no effect on crude fat content (McEwen and Dietz 1965).

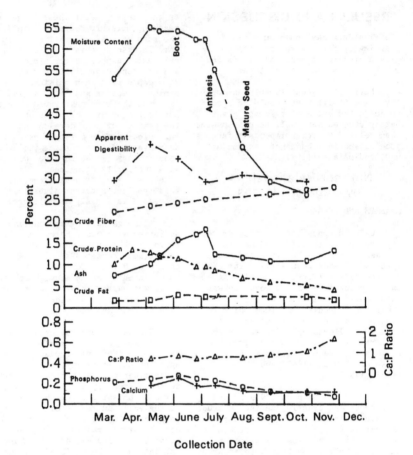

Figure 2.--Seasonal trends of constituents in Idaho fescue (Pinus ponderosa/Purshia tridentata/Festuca idahoensis).

6

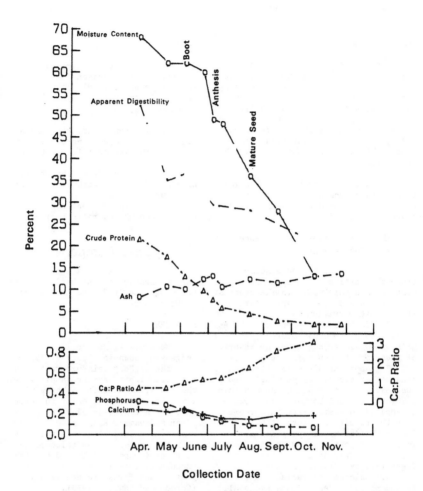

Figure 3.--*Seasonal trends of constituents in bluebunch wheatgrass*
(Artemisia arbuscula/Agropyron spicatum).

SHRUBS

Seasonal trends for phosphorus, ash, crude protein, and moisture showed association for shrub species reaching a peak during spring and a low point in fall or winter (figs. 4 and 5). Studies in Virginia exhibited the same pattern (Hundley 1959). Dietz et al. (1962) described similar trends for crude protein and phosphorus in shrubs in Colorado.

The calcium:phosphorus ratio for shrubs was narrow when moisture content was high and widest when moisture content was low.

There were several differences between shrubs and grasses:

1. Crude fat showed a seasonal trend the reverse of phosphorus, ash, protein, and moisture. In the grasses, crude fat had no consistent relationship to other nutrients.
2. The seasonal trend of ash was similar to phosphorus, protein, and moisture. Ash trends were the reverse of these constituents in the grasses.
3. Apparent digestibility and crude fiber trends varied widely among species. In bitter brush and mountain mahogany, crude fiber was the reverse of apparent digestibility in agreement with the results of Dietz et al. (1962). Sagebrush (*Artemisia* spp.) did not follow this pattern, and crude fiber curves were unrelated or similar to digestibility curves.
4. Calcium was variable and showed no consistent trends among all shrub species.
5. Growth stages varied in relationship to moisture and other nutrients in shrubs. For example, bitter brush and mountain mahogany flowered in spring while moisture was increasing and before maximum water content was reached, and the sagebrushes flowered in fall while moisture was lowest. In contrast, grasses flowered after the period of maximum water content while moisture was decreasing.

FORBS

The number of samples was inadequate to show complete seasonal trends in nutrients in forbs (fig. 6) which were available for grazing for only a short period. Despite this, data show that

1. Moisture content remained high during the sampling period (the period forbs are most likely to be grazed).
2. Apparent digestibility was high and decreased with maturity as did crude protein.
3. Calcium content and the calcium:phosphorus ratio increased with maturity.

Species Comparisons

Individual species are compared at various seasons on the basis of their phenology (table 2) and content of specific nutrients (below).

MOISTURE

Moisture trends for grasses and sedge are shown in figure 7. Species with the highest moisture content at the beginning of spring sampling were the lowest at maturity. Those with a low moisture content early in spring ended with a high level at maturity. After midsummer, Ross' sedge (*Carex Rossii*) remained higher in moisture than the grasses. The value of sedges as range forage is enhanced by high succulence late in the season when grasses and forbs are nearly dry (Gordon and Sampson 1939).

A generalized moisture curve is presented in figure 8. This curve summarizes the data of figure 7 for four grasses.

Note that moisture content is high but is decreasing slightly at the boot stage. During anthesis, moisture begins a rapid decline which continues through seed formation and seed maturity.

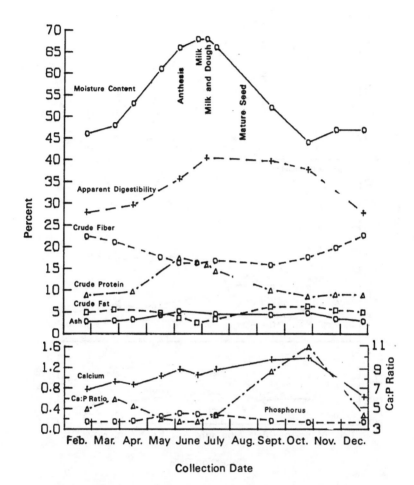

Figure 4.--Seasonal trends of constituents in bitter brush (Pinus ponderosa/Purshia tridentata/Festuca idahoensis).

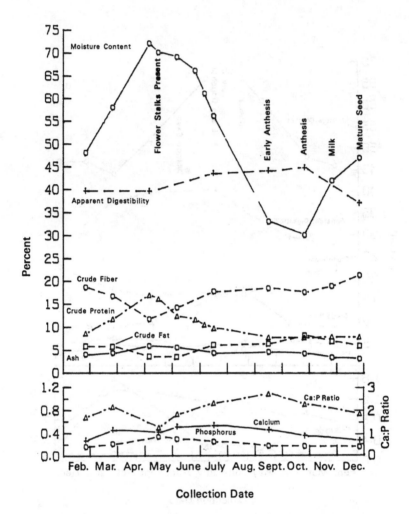

Figure 5.--*Seasonal trends of constituents in scabland sagebrush
(Artemisia arbuscula/Poa secunda).*

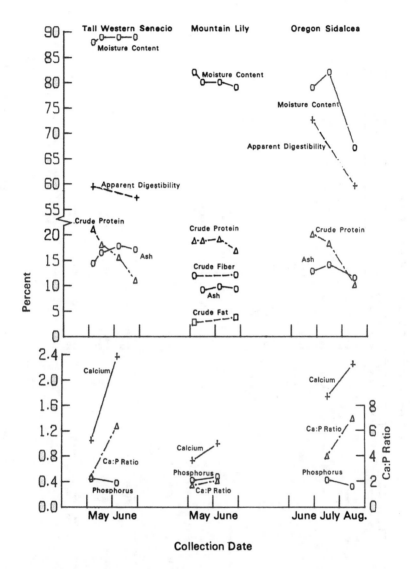

Figure 6.--Seasonal trends of constituents in tall western senecio, mountain lily, and Oregon sidalcea.

Table 2.-- *Average growth stage of species at time of collection*

Species	Mid-April	Early May	Mid-May	Late May	Early June	Late June	Early July	Mid-July	Mid-August	Mid-Sept.	Late Oct.	Late Nov.	Late Dec.
Grasses													
Sandberg bluegrass	boot			late anthesis		mature seed		dry					
Ross' sedge[1]			anthesis		dough	late dough	mature seed?						
Thurber needlegrass		early boot			in head	early milk	late milk	late dough -mature seed	mature seed				
Squirreltail					boot	anthesis	early milk	dough	mature seed				
Junegrass					boot		anthesis						
Bluebunch wheatgrass					boot		anthesis	late dough -mature seed	mature seed				
Idaho fescue					boot		anthesis	dough	mature seed				
Forbs													
Mountain lily		anthesis	late anthesis		milk	dough							
Upland larkspur			anthesis										
Tall western senecio		flower stalks	buds closed		anthesis	post-anthesis		mature seed					
Oregon sidalcea						buds closed	anthesis		mature seed				
Shrubs													
Mountain mahogany	few buds	many buds	anthesis			late milk	dough	dough	mature seed				
Bitter brush					anthesis	milk	milk-dough	anthesis-milk	mature seed		leaves brittle		
Scabland sagebrush			flower stalks		stalks enlarged					early anthesis	anthesis	anthesis-milk	mature seed
Big sagebrush			vegetative growth		flower stalks	flower stalks	stalks enlarged			early anthesis	anthesis	anthesis-milk	mature seed

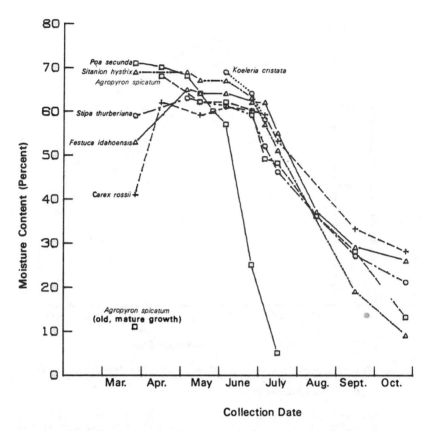

Figure 7.--Seasonal moisture trends in herbaceous species.

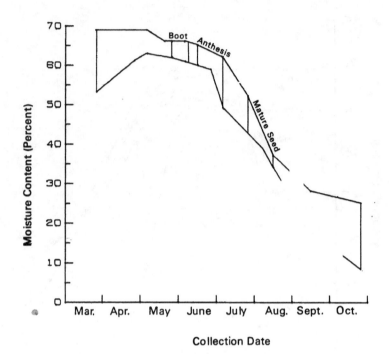

Figure 8.--Moisture curve which summarized the data presented in
figure 7 for Idaho fescue, Thurber needlegrass, bluebunch wheatgrass, and squirreltail.

Each shrub had a distinct moisture curve (fig. 9).

Forbs were higher in moisture
content than grasses or shrubs and
showed little change during sampling,
but dried rapidly after maturity.

CRUDE PROTEIN

All grasses were high in crude
protein during the early spring
(fig. 10). However, Sandberg bluegrass, squirreltail, and bluebunch
wheatgrass were higher than the other
species at the earliest sampling dates
when deer probably make considerable
use of green grass. Idaho fescue contained about half as much protein as
these species during this period.

Bluebunch wheatgrass had the
lowest crude protein content among the
grasses from flowering (in July) through
maturity. Thurber needlegrass had the
highest crude protein content among
the grasses during this period.

Ross' sedge contained less protein
than the grasses in March and April.
Afterwards, it followed a trend similar
to that of Idaho fescue but ended the
season higher in protein than any of
the grasses.

Those grasses with lowest protein
levels in early spring had the highest
levels in the fall. This same relationship prevailed for moisture content.

14

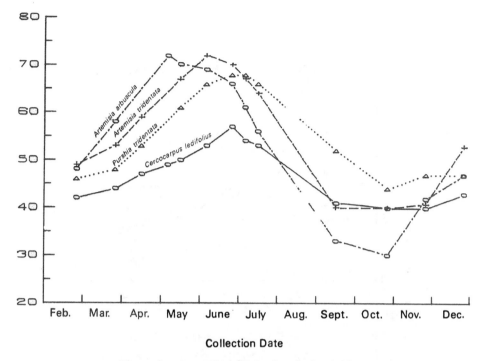

Figure 9.--Seasonal moisture trends in shrubs.

Urness (1966) determined that crude protein levels during the winter exceeded 10 percent for several range grasses growing near Silver Lake. Squirreltail, bluebunch wheatgrass, and Sandberg bluegrass were high (18-25 percent), while winter growth of Idaho fescue was lower (10-14 percent) with 8-10 percent for Ross' sedge.

Skovlin (1967) reported different results for summer collections of Idaho fescue and bluebunch wheatgrass in the Blue Mountains. Protein trends for these species were nearly parallel, and wheatgrass averaged 1 percent more than fescue. Wheatgrass contained 2 to 3 percent less protein than fescue during summer in the Silverlake study.

Similar to the results of Urness (1966), browse species ranged from 8 to 10 percent crude protein during the winter period. During the spring, protein levels were higher but varied considerably between species. By late summer, crude protein dropped to the 8- to 10-percent level, where it remained the rest of the season.

Seasonal variation in protein (difference between highest and lowest level) for grass species ranges from 9 percent in Idaho fescue to 20 percent in bluebunch wheatgrass. Variation within the shrubs was much less. Scabland sagebrush had the greatest variation (9 percent). Mountain mahogany varied least (2.6 percent).

15

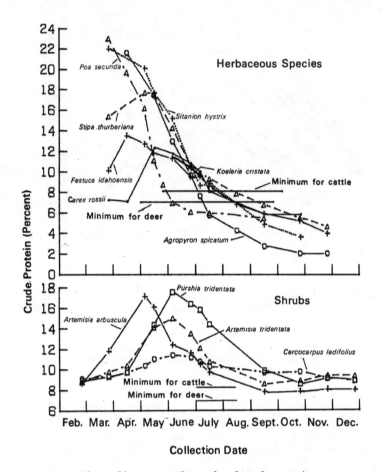

Figure 10.--Seasonal trends of crude protein.

Forbs were higher in crude protein than either grasses or browse species during the sampling. Mountain lily retained most of its protein, but others tended to drop in protein as they matured.

Whitman et al. (1951) suggested crude protein levels of 8.3 percent in range forage would meet the requirements of 900- to 1,100-pound nursing cows, indicating green grass met the protein requirement of cattle during the livestock grazing season used during the study.

Sandberg bluegrass and bluebunch wheatgrass became deficient in protein during anthesis. Idaho fescue, Thurber needlegrass, and squirreltail became deficient sometime during the early or middle stages of seed formation.

Ross' sedge exceeded the protein requirement only during May, June, and early July. Forbs were adequate during the short period available for grazing.

The shrubs (fig. 10) contained adequate protein throughout the study, except for a slight deficiency in Scabland sagebrush from September through January. Shrubs contained more protein than grasses throughout summer and fall. They would be an important protein source to cattle if grazing occurred during this period. Cook and Harris (1968) observed that browse consumption by cattle increased as the summer grazing season progressed in Utah. This late browsing tendency, along with the high protein content of shrubs at this time, suggests limiting late season grazing to browse ranges. The practicality of using protein supplements on nonbrowse ranges, as well as potential wildlife-livestock conflicts on browse ranges, may make this alternative less desirable.

The minimum level of crude protein required in the diet for deer is 6 to 7 percent (Dietz 1965). At a minimum requirement of 7 percent, Sandberg bluegrass became deficient in early June and bluebunch wheatgrass in early July. The other grasses, including Ross' sedge, would be satisfactory until later in the summer. The shrubs met the minimum requirement for protein throughout the entire year.

ASH (TOTAL MINERAL MATTER)

Herbaceous species were higher in ash content than shrubs at all sampling dates (fig. 11). They ranged between 7 and 18 percent, depending on species and development stage. There was considerable variation among and within species which was difficult to interpret.

Woody species ranged from 2 to 6 percent ash throughout the season (fig. 11) which is similar to results reported by Dietz et al. (1962). All species had similar but distince trends, and mountain mahogany exhibited the least seasonal variation.

CALCIUM

The grasses (and Ross' sedge) ranged between 0.1 and 0.4 percent calcium throughout the sampling period (fig. 12). Sandberg bluegrass had the highest calcium content during the early spring period, with Ross' sedge lowest. The sedge was higher than the grasses by anthesis (May) and remained high most of the summer and fall. Idaho fescue had the lowest calcium content among the grasses during the late summer and fall period.

Shrubs contained more calcium than grasses and exhibited greater variation among species (fig. 12). Scabland sagebrush had the lowest calcium content of the shrubs. It was closely paralleled by big sagebrush which remained slightly higher. Mountain mahogany was the highest of the shrubs in calcium except during the early fall when it was surpassed by bitter brush.

Three forbs were analyzed for

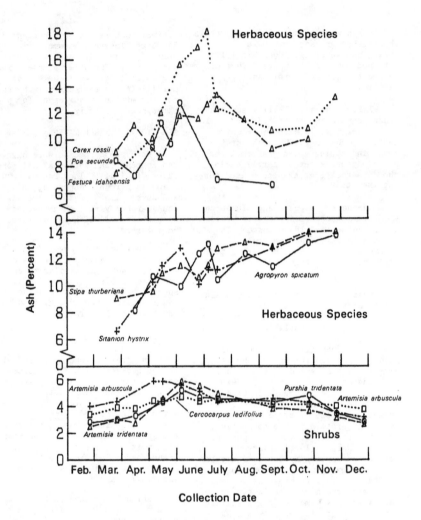

Figure 11.--Seasonal trends of ash (total mineral matter).

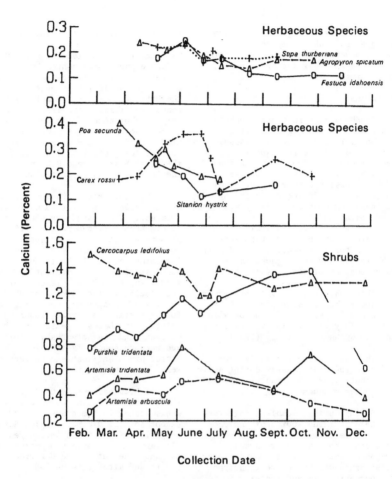

Figure 12.--Seasonal trends of calcium.

19

calcium. All were higher in calcium than grasses. When compared with shrubs, mountain lily remained intermediate between big sagebrush and bitter brush.

PHOSPHORUS

Sandberg bluegrass and bluebunch wheatgrass had the highest spring phosphorus levels among the grasses (fig. 13) but decreased rapidly in these species and dropped below the others by early summer. Ross' sedge had the lowest phosphorus content of species sampled during early spring but increased substantially by mid-May. Idaho fescue had less phosphorus than the other grasses during early spring leaf development but more than Ross' sedge. After boot state (early June), Idaho fescue had the highest phosphorus content of the grasses.

These phosphorus levels are generally lower than grasses studied by Gordon and Sampson (1939) and Skovlin (1967). They are similar to results reported by Whitman et al. (1951).

The shrubs were generally lower in phosphorus than grasses during early spring but higher than the grasses and sedge after midsummer (fig. 13).

During spring, scabland sagebrush was the best shrub source of phosphorus. Big sagebrush contained more phosphorus than the other shrubs throughout the summer, fall, and winter. Bitter brush was the poorest phosphorus source in fall and winter. Mountain mahogany was intermediate and less variable in phosphorus content.

Similar results were obtained by Dietz et al. (1962) for bitter brush, big sagebrush, and Colorado mountain mahogany (*Cercocarpus montanus*) in Colorado and by Smith (1963) for big sagebrush near Silver Lake.

Three forbs were analyzed for phosphorus and were higher than either grasses or shrubs during the period

they were available for sampling.

According to Black et al. (1943), a forage phosphorus content of 0.13 percent is necessary to meet the minimum range cattle requirements. However, the National Research Council (1958) considered forage with less than 0.15 percent phosphorus deficient. Still higher phosphorus levels were recommended by Cook and Harris (1968)--0.17 percent for gestation, 0.22 percent for the first 8 weeks, and 0.20 percent for the last 12 weeks of lactation.

At 0.15 percent, phosphorus became deficient in grasses at dates ranging from early June (Sandberg bluegrass) to mid-August (Idaho fescue) (fig. 13). When related to growth stages, bluebunch wheatgrass became deficient during anthesis, and Idaho fescue and Ross' sedge became deficient soon after seed maturity. Other grasses were intermediate. The forbs were good sources of phosphorus, supplying two to three times the amount required by cattle. Two browse species, mountain mahogany and big sagebrush, supplied adequate phosphorus season long, but bitter brush and·scabland sagebrush were deficient in fall.

If 0.20 percent is used, phosphorus becomes deficient about 3 to 4 weeks sooner in the grasses. And, except for the mid-May sample, Ross' sedge is deficient the entire year. All shrubs have ample phosphorus during May and June but become deficient before and after this period.

Regardless of the minimum requirement assumed, the data indicate that phosphorus is likely to be deficient in cattle diets during late summer and fall. Consequently, supplementation may be necessary where this range is used for late season grazing.

Minimum phosphorus requirements have not been determined for deer, but 0.16 percent is minimal for pregnant ewes (Dietz 1965). If deer are similar, grass was deficient in the late spring

20

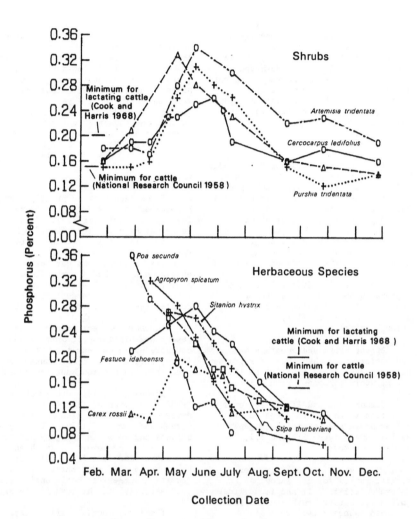

Figure 13.--Seasonal trends of phosphorus.

(Sandberg bluegrass) or summer (all other grasses). Ross' sedge was deficient during the early spring, summer, and fall. Scabland sagebrush and bitter brush were phosphorus deficient during the fall (also winter for bitter brush), while mountain mahogany and big sagebrush were satisfactory throughout the year.

CALCIUM:PHOSPHORUS RATIO

The grasses, including Ross' sedge, had calcium:phosphorus ratios between 0.5:1 and 3:1 (fig. 14). Each species, except the sedge, had a ratio near 1:1 early in the season, widening as the season progressed, with greatest increases in Sandberg bluegrass (2:1 by mid-July) and bluebunch wheatgrass (3:1 in October). Little change occurred in Idaho fescue until midfall.

These results differ from those reported by Gordon and Sampson (1939) who found that in California foothill grasses, this ratio (1:1) was constant throughout the season.

Bitter brush and mountain mahogany had ratios wider than those for grasses. Bitter brush ranged from 4:1 to 11:1 and mountain mahogany, from 5:1 to 9:1.

Sagebrushes had little seasonal variation in the ratio compared with bitter brush and mountain mahogany. Ratios were between 1:1 and 3:1, similar to grasses.

Dietz et al. (1962) reported similar ratios for big sagebrush in Colorado (2:1 to 3:1) and found wider values for bitter brush and Colorado mountain mahogany (4:1 to 8:1).

The ratio varied between three forbs. Tall western senecio and Oregon sidalcea ranged between 2:1 and 7:1, while mountain lily remained near 2:1.

According to the recommendation of 0.5:1 to 2:1 as a standard for cattle (Maynard and Loosli 1962),

scabland sagebrush provided the most desirable balance of these minerals in the shrubs. Big sagebrush was slightly higher than this during part of the season. Both mountain mahogany and bitter brush had wider ratios than recommended the entire year. The grasses and sedge were satisfactory most of the season except for the bluebunch wheatgrass during September and October. Mountain lily provided the lowest and most desirable ratio.

CRUDE FIBER

Crude fiber varied greatly among herbaceous species during early spring leaf development (fig. 15). Ross' sedge was high in crude fiber in March (26 percent) but lower than the other species later in the season. Although Sandberg bluegrass was earliest in development, it had a curve similar to squirreltail. Both were low in crude fiber in March (14 percent) but had doubled their fiber content by late June.

Thurber needlegrass and Idaho fescue had similar post-July trends but were different earlier. Idaho fescue increased in fiber content at a constant rate throughout the season.

Shrubs (fig. 15) were lower in crude fiber than grasses (except during early spring) as was also reported by Gordon and Sampson (1939). The sagebrushes and mountain mahogany had similar trends--low in midspring and high in the summer and fall (fig. 15). Bitter brush had a different curve that was lowest in summer (anthesis until after seed maturity) and highest in winter.

The forb, mountain lily, contained 12 percent crude fiber; and except for a May sample of scabland sagebrush, it was the lowest in crude fiber of all species.

Dietz et al. (1962) reported similar trends but higher values for crude fiber in bitter brush and scabland sagebrush and a quite different trend for Colorado mountain mahogany,

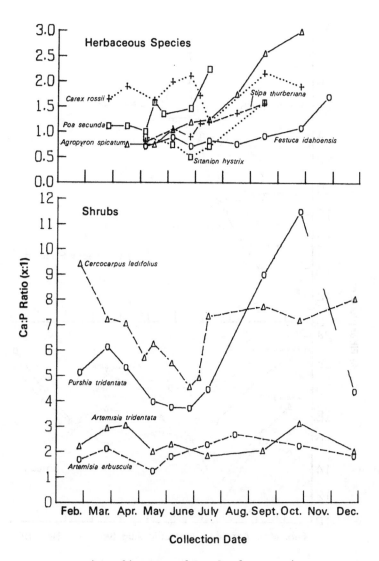

Figure 14.--Seasonal trends of Ca:P ratio.

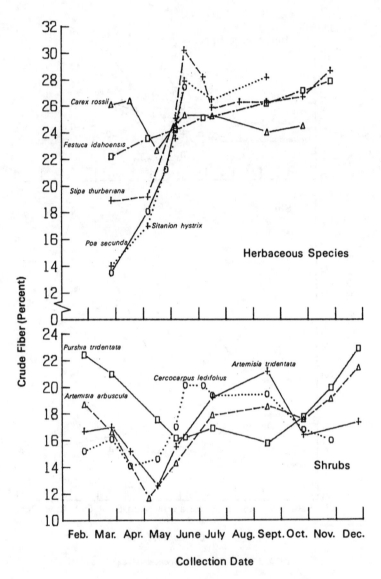

Figure 15.--Seasonal trends of crude fiber.

as well as higher values for *C. ledifolius*. Smith (1963) reported a similar seasonal trend for big sagebrush which was slightly higher than results reported here.

CRUDE FAT (ETHER EXTRACT)

The grasses ranged from 1 to 4 percent crude fat (fig. 16). In the early spring needlegrass, squirreltail, and bluegrass were better fat sources than Idaho fescue or Ross' sedge. Crude fat was not determined for bluebunch wheatgrass.

The shrubs were better sources of fat than grasses, especially during winter and early spring when deer are greatly dependent on browse (fig. 16). Big sagebrush was the highest of the shrubs in crude fat during this period. Mountain mahogany was nearly as high and retained a high level of crude fat longer in the spring. Bitter brush generally contained less crude fat than other shrubs.

Dietz et al. (1962) also found bitter brush low in crude fat content and reported 12 to 15 percent crude fat for big sagebrush--much higher than the results reported here. Smith (1963) found seasonal trends for big sagebrush similar to those in this study but reported higher values. The seasonal trend reported here for big sagebrush appears to be accurate but absolute values are unexplainably low.

Crude fat was determined for one forb, mountain lily, and fat content remained between 2 and 4 percent.

APPARENT DIGESTIBILITY

Herbaceous species exhibited downward trends in dry matter disappearance (DMD, 24-hour digestion period) during most of the growing season (fig. 17), with absolute values lower than those reported for grasses by Wallace and Raleigh (1962), Wallace et al. (1965), and Pearson (1964).

Grasses (except for Idaho fescue) were highly digestible early in the season (44- to 67-percent DMD) but decreased to values less than 30 percent. Idaho fescue was low in digestibility (30 percent) at the earliest collection date. In the fall, it was higher in digestibility than other grasses. Sandberg bluegrass dropped more rapidly in DMD than other grasses, but this was related to its short growth period and early maturity. Ross' sedge, although less digestible than most grasses in the spring, remained well above 40-percent DMD at the summer and fall collections. This may be related to its tendency to remain succulent while the grasses mature and dry.

The shrubs had less seasonal variation in apparent digestibility than most herbaceous species (fig. 17). They were less digestible than most grasses early in the season but higher than grasses late in the season. Big sagebrush was consistently higher in DMD than other shrubs and was closely paralleled by scabland sagebrush at a lower level. Mountain mahogany and bitter brush had more variation in DMD with growth stage than did the sagebrush species. During late spring and summer, mountain mahogany was the lowest shrub in digestibility; and during the winter and early spring, bitter brush was least digestible.

Only two forbs were studied. Both were very high in DMD (57 to 73 percent) during the short period they were sampled.

Digestibility was also determined with a 48-hour digestion period for two species, bitter brush and Idaho fescue. When compared with the 24-hour digestion period there was a sizable increase in DMD for the grass, but only a small increase (except in October) for the shrub. Wallace et al. (1965) reported that, for meadow and rye hay, digestibility by sheep was underestimated by the 24-hour period and overestimated by the 48-hour period, suggesting that the best approximation would be intermediate.

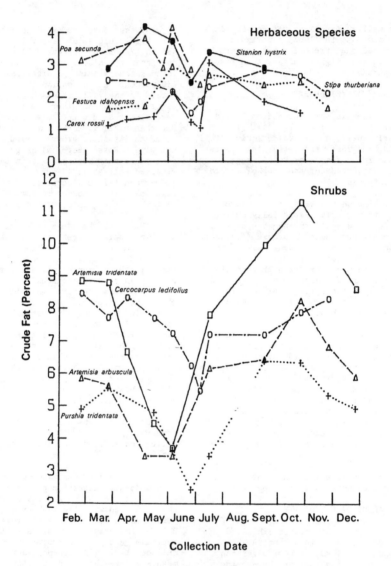

Figure 16.--Seasonal trends of crude fat.

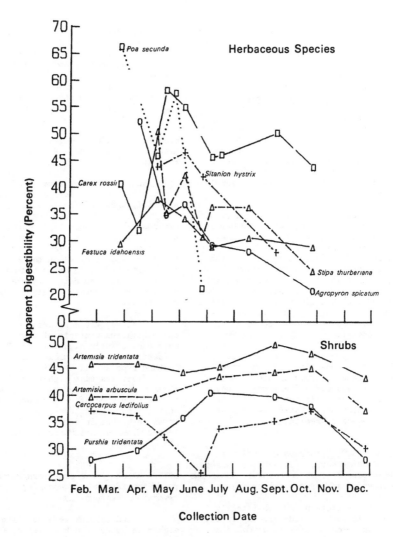

Figure 17.--Seasonal trends of apparent digestibility (24-hour digestion period).

Because these data were obtained using rumen juice from a steer, it is not known whether the results are directly applicable to deer.

The Effect of Site on Chemical Composition

Chemical composition varied with the site or vegetation type from which the species was collected. Some differences might be attributed to the sampling vagaries. However, some variation was the result of differences in the growth stages of plants sampled in the various types.

Other factors can also contribute to such differences. Cook and Harris (1950b) attributed the change in plant composition by sites to differences in shade and soil moisture. Also, the soil character may affect plant absorption and vegetative composition (Richardson et al. 1954). Nutrient content depends, within limits, on interaction between chemical, physical, and biological factors associated with the environment.

The apparent effects of site on chemical composition are summarized for those constituents which had consistent seasonal differences between sites. These differences cannot be substantiated statistically because of the lack of replicates.

IDAHO FESCUE (*Pinus ponderosa/Purshia tridentata/Festuca idahoensis*; *Purshia tridentata/Festuca idahoensis*; *Juniperus occidentalis/Festuca idahoensis*)

Crude protein was consistently higher in Idaho fescue growing in the *Purshia tridentata/Festuca idahoensis* type than in the *Pinus ponderosa/ Purshia tridentata/Festuca idahoensis* or *Juniperus occidentalis/Festuca idahoensis* types during May through September. Ash was highest in the *Purshia tridentata/Festuca idahoensis* type and lowest in the *Juniperus occidentalis/Festuca idahoensis* type from

July until the end of the season. Crude fat content of Idaho fescue was highest in the *Purshia tridentata/ Festuca idahoensis* type throughout the season. Except for one sampling date, crude fat remained the lowest in the *Juniperus occidentalis/Festuca idahoensis* type.

SQUIRRELTAIL (*Purshia tridentata-Artemisia arbuscula/Poa secunda*; *Artemisia arbuscula/ Poa secunda*)

Phenological development of squirreltail was 2 weeks earlier in *Artemisia arbuscula/Poa secunda* than the *Purshia tridentata-Artemisia arbuscula/Poa secunda* type. Squirreltail plants were also small and stunted in *Artemisia arbuscula/Poa secunda* type but large and vigorous in the *Purshia tridentata-Artemisia arbuscula/Poa secunda* type. The data reflected the differences in growth stages for nearly all constituents.

Moisture, crude protein, crude fat, and phosphorus contents of squirreltail were consistently higher in the *Purshia tridentata-Artemisia arbuscula/Poa secunda* type at all sampling dates. Crude fiber was also higher in the *Purshia tridentata-Artemisia arbuscula/ Poa secunda* type after mid-June. Except for one sample date, ash content remained higher in squirreltail collected from the *Artemisia arbuscula/ Poa secunda* type.

BITTER BRUSH (*Pinus ponderosa/Purshia tridentata/Festuca idahoensis*; *Purshia tridentata/Festuca idahoensis*)

Ash and calcium were consistently higher in bitter brush collected from *Pinus ponderosa/Purshia tridentata/ Festuca idahoensis* than from the *Purshia tridentata/Festuca idahoensis* type. Except for the spring period, phosphorus content was also higher in the *Pinus ponderosa/Purshia tridentata/ Festuca idahoensis* type. Crude fiber was much higher in bitter brush from the *Pinus ponderosa/Purshia tridentata/ Festuca idahoensis* type from May through October.

SUMMARY

Seasonal nutrient trends were determined for several prominent grasses, forbs, and shrubs found in a portion of the Silver Lake winter deer range about 80 miles south of Bend, Oregon. Sampling included one complete annual growth cycle for most species. Composite samples (each observation) were collected from within particular vegetation types, and the average growth stage was estimated for each sample. Laboratory analyses were made for moisture, crude protein, ash, calcium, phosphorus, crude fiber, crude fat (ether extract), and apparent digestibility (dry matter disappearance).

In the grasses, calcium, phosphorus, crude protein, apparent digestibility, and moisture generally declined as the season progressed. Crude fiber and ash content increased with plant maturity. Crude fat had no distinct trend that was related to plant development. The calcium:phosphorus ratio widened in most species as the season progressed.

Seasonal trends for phosphorus, ash, crude protein, and moisture appear to be associated in all the shrubs studied. These constituents reached a peak during the spring and a low point in the fall or winter. Unlike in the grasses, crude fat in the shrubs had a distinct seasonal trend. This trend appeared to be inversely related to moisture content. Calcium, apparent digestibility, and crude fiber trends varied considerably among species.

The number of forb samples was inadequate to make positive statements about seasonal nutrient trends in these species. However, both moisture and apparent digestibility were high in forbs during the sampling period. The calcium:phosphorus ratio appeared to widen with plant maturity, whereas moisture, crude protein, and apparent digestibility tended to decrease.

The effect of site (vegetation type) on chemical composition was studied in bitter brush, Idaho fescue, and squirreltail. Probable differences between sites, found for several constituents, are considered to be the result of growth stage differences, shading, or other factors associated with the various environments sampled.

LITERATURE CITED

Association of Official Agricultural Chemists
    1965. Official methods of analysis. 10th ed., 957 p. Washington, D.C.

Beath, O. A., and J. W. Hamilton
    1952. Chemical composition of Wyoming forage plants. Wyo. Agric. Exp. Stn. Bull. 311, 40 p.

Bedell, T. E.
    1966. Seasonal cattle and sheep diets on *Festuca arundinacea-Trifolium subterraneum* and *Lolium perenne-Trifolium subterraneum* pastures in western Oregon. Ph.D. thesis, 283 p. Oreg. State Univ., Corvallis.

Black, W. H., L. H. Tash, J. M. Jones, and R. J. Kleberg, Jr.
    1943. Effects of phosphorus supplements on cattle grazing on range deficient in this mineral. U.S. Dep. Agric. Tech. Bull. 856, 23 p.

Cook, C. W., and L. E. Harris
    1950a. The nutritive content of the grazing sheep's diet on summer and
        winter ranges of Utah. Utah State Agric. Exp. Stn. Bull. 342, 66 p.

_____ and L. E. Harris
    1950b. The nutritive value of range forage as affected by vegetation type,
        site and state of maturity. Utah Agric. Exp. Stn. Tech. Bull. 344,
        45 p.

_____ and L. E. Harris
    1968. Nutritive value of seasonal ranges. Utah Agric. Exp. Stn.
        Bull. 472, 55 p.

Dealy, J. Edward
    1966. Bitterbrush nutritional levels under natural and thinned ponderosa
        pine. USDA For. Serv. Res. Note PNW-33, 6 p., illus. Pac. Northwest
        For. & Range Exp. Stn., Portland, Oreg.

Dietz, Donald R.
    1965. Deer nutrition research in range management. 13th North Am. Wildl.
        & Nat. Resour. Conf. Trans: 274-285.

_____ R. H. Udall, and L. E. Yeager
    1962. Chemical composition and digestibility by mule deer of selected
        forage species, Cache La Poudre Range, Colorado. Colo. Dep. Game
        & Fish Tech. Publ. 14, 89 p.

Gordon, A., and A. W. Sampson
    1939. Composition of common California foothill plants as a factor in range
        management. Calif. Agric. Exp. Stn. Bull. 627, 95 p.

Hitchcock, A. S.
    1950. Manual of the grasses of the United States. 2d ed., 1051 p.
        U.S. Dep. Agric. Misc. Publ. No. 200.

Hundley, L. R.
    1959. Available nutrients in selected deer browse species growing on
        different soils. J. Wildl. Manage. 23(1): 81-90.

Jackson, M. L.
    1960. Soil chemical analysis. 498 p. Englewood Cliffs, N.J.: Prentice-
        Hall, Inc.

Johnsgard, G. A.
    1963. Temperature and the water balance for Oregon weather stations. Oreg.
        State Univ. Agric. Exp. Stn. Spec. Rep. 150, 127 p. Corvallis.

Maynard, L. A., and J. K. Loosli
    1962. Animal nutrition. 5th ed. New York: McGraw-Hill Book Co., Inc.

McEwen, L. C., and D. R. Dietz
    1965. Shade effects on chemical composition of herbage in the Black Hills.
        J. Range Manage. 18(4): 184-189.

National Research Council
    1958. Nutrient requirements of domestic animals. IV. Nutrient requirements
          of beef cattle. Nat. Acad. Sci. Publ. 579, 32 p.

Pearson, Henry A.
    1965. Studies of forage digestibility under ponderosa pine stands. Soc.
          Am. For. Meet. Proc. 1964: 71-73.

Peck, M. E.
    1961. A manual of higher plants in Oregon. 2d ed., 936 p. Portland,
          Oreg.: Binfords and Mort.

Raleigh, R. J., and J. D. Wallace
    1965. Nutritive value of range forage and its effect on animal performance.
          Oreg. Agric. Exp. Stn. Spec. Rep. 189, 14 p.

Richardson, L. R., M. Speirs, and W. J. Peterson
    1954. Influence of environment on the chemical composition of plants.
          I. A review of the literature. South. Coop. Ser. Bull. 36: 1-5.

Skovlin, Jon M.
    1967. Fluctuations in forage quality on summer range in the Blue Mountains.
          USDA For. Serv. Res. Pap. PNW-44, 20 p. Pac. Northwest For. & Range
          Exp. Stn., Portland, Oreg.

Smith, G. E.
    1963. Nutritional effects of big sagebrush (*Artemisia tridentata* Nutt.)
          on deer. Master's thesis, 111 p. Oreg. State Univ., Corvallis.

Urness, P. J.
    1966. Influence of range improvement on composition production and utili-
          zation of deer winter range in central Oregon. Ph.D. thesis, 165 p.
          Oreg. State Univ., Corvallis.

Vallentine, J. R., and V. A. Young
    1959. Factors affecting the chemical composition of range forage plants
          on the Edwards Plateau. Texas Agric. Exp. Stn. MP-384, 8 p.

Wallace, J. D., and R. J. Raleigh
    1962. Effect of different levels of salt in a cottonseed meal supplement
          for yearling cattle on crested wheatgrass pasture. Oreg. Agric.
          Exp. Stn. Tech. Pap. 1556, 5 p.

_____, C. B. Rumburg, and R. J. Raleigh
    1961. Evaluation of range and meadow forages at various stages of maturity
          and levels of nitrogen fertilization. West. Sec. Am. Soc. Anim.
          Prod. Proc. 12: 1-6.

_____, C. B. Rumburg, and R. J. Raleigh
    1965. A comparison of *in vitro* techniques and their relation to *in vivo*
          values. West. Sec. Am. Soc. Anim. Sci. Proc. 16: 1-6.

Whitman, W. C., D. W. Bolin, E. W. Kosterman, et al.
    1951. Carotene, protein and phosphorus in range and tame grasses of
           western North Dakota. Agric. Exp. Stn., N. D. Agric. Coll. Bull.
           370, 55 p.

Willis, J. B.
    1960. The determination of metals in blood serum by an atomic absorption
           spectroscopy. I. Calcium. Spectrochim. Acta 16: 259-272.

32

☆ U S. GOVERNMENT PRINTING OFFICE 1975—698-522 /

The mission of the PACIFIC NORTHWEST FOREST AND RANGE EXPERIMENT STATION is to provide the knowledge, technology, and alternatives for present and future protection, management, and use of forest, range, and related environments.

Within this overall mission, the Station conducts and stimulates research to facilitate and to accelerate progress toward the following goals:

1. Providing safe and efficient technology for inventory, protection, and use of resources.

2. Development and evaluation of alternative methods and levels of resource management.

3. Achievement of optimum sustained resource productivity consistent with maintaining a high quality forest environment.

The area of research encompasses Oregon, Washington, Alaska, and, in some cases, California, Hawaii, the Western States, and the Nation. Results of the research will be made available promptly. Project headquarters are at:

Fairbanks, Alaska   Portland, Oregon
Juneau, Alaska   Olympia, Washington
Bend, Oregon   Seattle, Washington
Corvallis, Oregon   Wenatchee, Washington
La Grande, Oregon

Mailing address: Pacific Northwest Forest and Range
Experiment Station
P.O. Box 3141
Portland, Oregon 97208

CPSIA information can be obtained
at www.ICGtesting.com
Printed in the USA
LVHW021253071118
596294LV00004B/613